# THE POETRY OF GADOLINIUM

# The Poetry of Gadolinium

Walter the Educator

Silent King Books a WhichHead Imprint

Copyright © 2024 by Walter the Educator

All rights reserved. No part of this book may be reproduced in any manner whatsoever without written permission except in the case of brief quotations embodied in critical articles and reviews.

First Printing, 2024

Disclaimer
This book is a literary work; poems are not about specific persons, locations, situations, and/or circumstances unless mentioned in a historical context. This book is for entertainment and informational purposes only. The author and publisher offer this information without warranties expressed or implied. No matter the grounds, neither the author nor the publisher will be accountable for any losses, injuries, or other damages caused by the reader's use of this book. The use of this book acknowledges an understanding and acceptance of this disclaimer.

"Earning a degree in chemistry changed my life!"
— Walter the Educator

dedicated to all the chemistry lovers, like myself, across the world

# CONTENTS

Dedication . . . . . . . . . . . . . v

Why I Created This Book? . . . . . . . . . 1

**One** - A Jewel Among Elements . . . . . . 2

**Two** - Infinite Arc . . . . . . . . . . 4

**Three** - Forever Intertwined . . . . . . . . 6

**Four** - Element Of Wonder . . . . . . . 8

**Five** - In Every Place . . . . . . . . . 10

**Six** - Symbol Of Strength . . . . . . . . 12

**Seven** - Gem So Fine . . . . . . . . . 14

**Eight** - So Opulent . . . . . . . . . . 16

**Nine** - Truly Astound . . . . . . . . . 18

**Ten** - Dazzling . . . . . . . . . . . 20

**Eleven** - Elegance And Embrace . . . . . . 22

**Twelve** - Like No Other . . . . . . . . 24

**Thirteen** - Bold And Refined . . . . . . . 26

**Fourteen** - For All To See . . . . . . . 28

**Fifteen** - Captivating . . . . . . . . . 30

**Sixteen** - Inspire . . . . . . . . . . . 32

**Seventeen** - Cherished And Defined . . . . 34

**Eighteen** - Admire Your Allure . . . . . . . 36

**Nineteen** - Guardian Of Health . . . . . . 38

**Twenty** - Atomic Might . . . . . . . . . 40

**Twenty-One** - Remarkable . . . . . . 42

**Twenty-Two** - Essence Of Grand . . . . . 44

**Twenty-Three** - Chemistry's Art . . . . . 46

**Twenty-Four** - Symbol Of Progress . . . . . 48

**Twenty-Five** - Atomic Structure . . . . . . 50

**Twenty-Six** - 64 . . . . . . . . . . 52

**Twenty-Seven** - Magnetic Glance . . . . 54

**Twenty-Eight** - Contrast Agent . . . . . . 56

**Twenty-Nine** - Celestial Light . . . . . . 57

**Thirty** - Treasure Of Science . . . . . . . 59

**Thirty-One** - Atomic Reign . . . . . . . . 61

**Thirty-Two** - Gadolinium, Orchestrating Atoms . . . . . . . . . . . . . . 62

**Thirty-Three** - Scientific Swirl . . . . . . . . . 64

**Thirty-Four** - Stand Unique . . . . . . . . . . 65

**Thirty-Five** - Fascination . . . . . . . . . . . 67

About The Author . . . . . . . . . . . . . . 69

# WHY I CREATED THIS BOOK?

Creating a poetry book about the chemical element of Gadolinium can offer a unique and intriguing perspective on science and art. Gadolinium, with its atomic number 64, possesses fascinating properties that can be explored through the medium of poetry. By combining scientific knowledge with the creative expression of poetry, this book can educate and entertain readers in a captivating way. It can shed light on the element's history, characteristics, and applications, while also allowing for imaginative interpretations and emotional connections. This book can bridge the gap between science and the arts, encouraging interdisciplinary thinking and fostering a deeper appreciation for both.

# ONE

# A JEWEL AMONG ELEMENTS

In the realm of elements, a gem does shine,
Gadolinium, a treasure so divine,
With atomic number sixty-four it stands,
A lustrous metal in enchanted lands.

Within its core, electrons dance in grace,
A symphony of particles in space,
Magnetic nature, a rare delight,
Captivating hearts both day and night.

Gadolinium, a marvel to behold,
Its properties, a story yet untold,
Paramagnetic prowess it possesses,
Attracting fields with mystical finesse.

Fluorescent hues, in luminescent glow,
A kaleidoscope of colors it does show,

From vibrant blues to shades of green,
A dazzling spectacle, a vibrant scene.
 In medical realms, it finds its worth,
As contrast agents for a clearer berth,
MRI machines, they benefit from,
Gadolinium's prowess, a breakthrough to come.
 Oh Gadolinium, element of charm,
In laboratories, your secrets alarm,
A testament to nature's grand design,
A jewel among elements, forever mine.

# TWO

# INFINITE ARC

In a realm of science, a tale unfolds,
Of a rare element, Gadolinium it beholds,
With atomic number sixty-four, it reigns supreme,
A luminescent gem, a poet's dream.

Within its structure, a magnetic might,
Gadolinium dances in ethereal light,
Paramagnetic essence, it proudly displays,
Drawing fields in curious arrays.

In the depths of labs, its secrets reside,
A catalyst of wonders, it cannot hide,
From medical marvels to quantum quests,
Gadolinium's touch, innovation manifests.

A chameleon of color, a mesmerizing hue,
Gadolinium adorns the periodic tableau,

From azure blues to emerald greens,
Its radiant palette, a sight to be seen.

In the land of medicine, it finds its role,
Enhancing diagnostics, a remarkable stroll,
MRI's ally, a contrast it brings,
Revealing hidden truths, life's intricate strings.

But beyond the lab, Gadolinium inspires,
A poet's muse, its essence transpires,
A symbol of curiosity, of boundless might,
Gadolinium's story, a beacon of light.

So let us marvel at this element rare,
Gadolinium, a jewel beyond compare,
Its scientific wonders, its creative spark,
A testament to nature's infinite arc.

# THREE

# FOREVER INTERTWINED

In the realm of elements, a mystery unfurls,
Gadolinium, a tale that transcends the world,
With atomic number sixty-four, it takes its stance,
A symphony of electrons, a cosmic dance.

A silvery metal, with luster so divine,
Gadolinium, a treasure that's hard to define,
Its magnetic charm, a captivating spell,
Drawing forces with grace, a tale to tell.

Within its core, a secret dwells,
Gadolinium, the enigma that compels,
Paramagnetic prowess, it proudly displays,
In the presence of fields, its spirit never fades.

Oh, Gadolinium, your luminescent glow,
Paints the universe with colors that flow,

From neon blues to verdant greens,
A celestial kaleidoscope, a sight unseen.

In the realm of medicine, you find your worth,
An ally in diagnosis, a treasure from Earth,
MRI's companion, a contrast that shines,
Unveiling hidden truths, where healing aligns.

But beyond the lab, your essence inspires,
A symbol of resilience, as the world transpires,
Gadolinium, you teach us to adapt and evolve,
To embrace the unknown, to revolve and resolve.

So let us celebrate this element of grace,
Gadolinium, a wonder we embrace,
A cosmic enigma, a muse for the mind,
In the tapestry of elements, forever intertwined.

# FOUR

# ELEMENT OF WONDER

In the realm of elements, a star does gleam,
Gadolinium, a luminary, a radiant beam,
With atomic number sixty-four, it claims its place,
A captivating metal, an ethereal grace.
  Within its heart, a magnetic allure,
Gadolinium's enchantment, pure and secure,
Paramagnetic essence, it proudly holds,
Drawing fields of wonder, as the story unfolds.
  Oh Gadolinium, your colors ignite,
A vibrant palette, a mesmerizing sight,
From turquoise blues to emerald greens,
You paint the universe in kaleidoscopic scenes.
  In the realm of science, your worth is known,
A key to innovation, a foundation stone,

MRI's partner, revealing secrets unseen,
Gadolinium's gift, a medical dream.

But beyond the lab, your spirit transcends,
A symbol of resilience that never bends,
Gadolinium, you teach us to adapt and grow,
To face the unknown, to let our spirits flow.

In the tapestry of elements, you shine bright,
Gadolinium, a beacon in the darkest night,
A testament to nature's boundless art,
An element of wonder, forever in our heart.

# FIVE

# IN EVERY PLACE

In the realm of elements, a jewel is found,
Gadolinium, with mysteries profound,
Atomic number sixty-four, its claim to fame,
A metal of wonder, a captivating name.

Within its core, magnetic secrets lie,
Paramagnetic charm, it can't deny,
Drawing fields of fascination, it does dare,
Gadolinium's essence, beyond compare.

Oh Gadolinium, your hues enchant,
A symphony of colors, vibrant and grand,
From sapphire blues to emerald greens,
A visual feast, a kaleidoscope of scenes.

In the realm of medicine, you make your mark,
A contrast agent, a luminescent spark,

MRI's ally, revealing the hidden,
Gadolinium's magic, a gift from the forbidden.

But beyond the lab, your essence takes flight,
A symbol of resilience, shining so bright,
Gadolinium, you teach us to adapt,
To face life's challenges, never trapped.

In the tapestry of elements, you thrive,
Gadolinium, a treasure to keep alive,
A testament to nature's infinite grace,
An element of wonder, in every place.

# SIX

# SYMBOL OF STRENGTH

In the realm of elements, a gem so rare,
Gadolinium, with mysteries to share,
A magnetic marvel, captivating minds,
A symphony of electrons, dancing in binds.

Oh Gadolinium, your properties allure,
A paramagnetic treasure, so pure,
From azure hues to shades of jade,
A prism of colors, an enchanting cascade.

In the world of medicine, you find your way,
Enhancing diagnostics, a guiding ray,
MRI's companion, a contrast supreme,
Revealing secrets, like a lucid dream.

But beyond the lab, your story transcends,
A symbol of strength, as the universe extends,
Gadolinium, you teach us to adapt,

To embrace the unknown, where possibilities are mapped.
In the tapestry of elements, you shine,
Gadolinium, a treasure so divine,
A testament to nature's infinite art,
An element of wonder, forever in our heart.

# SEVEN

# GEM SO FINE

Gadolinium, element of wondrous might,
In the realm of chemistry, a radiant light,
From cerulean blue to verdant green,
Your luminescent beauty, a sight unseen.

In medicine's embrace, you play a vital role,
Enhancing imaging, a gift to behold,
MRI's ally, a contrast sublime,
Unveiling hidden truths, standing the test of time.

Yet beyond the lab, your story unfolds,
A symbol of resilience, as life beholds,
Gadolinium, you teach us to adapt,
To navigate challenges, never trapped.

In the tapestry of elements, you shine,
Gadolinium, a gem so fine,

A testament to nature's creative design,
An element of wonder, forever intertwined.
 So let us celebrate this element rare,
Gadolinium, with all its flair,
From scientific marvels to poetic verse,
You captivate us, from start to traverse.

# EIGHT

## SO OPULENT

Oh, Gadolinium, element of grace,
With a radiant glow, you embrace,
From cerulean blues to vibrant greens,
A captivating sight, like no other scenes.

    In the realm of science, you hold the key,
A magnetic marvel, so endlessly,
MRI's partner, a contrast supreme,
Revealing mysteries, like a vivid dream.

    But beyond the lab, your essence prevails,
A symbol of resilience that never fails,
Gadolinium, you teach us to endure,
To face life's challenges, to be secure.

    In the tapestry of elements, you shine,
Gadolinium, a treasure so divine,

A testament to nature's infinite art,
An element of wonder, from the very start.
 So let us honor you, dear Gadolinium,
With your luminescence, so opulent,
A beacon of light, in the darkest night,
Forever in our hearts, shining so bright.

# NINE

# TRULY ASTOUND

In the realm of elements, a jewel unfurled,
Gadolinium, a gem that lights up the world,
From cerulean depths to vibrant greens,
A symphony of colors, a celestial scene.

In the realm of medicine, you play a part,
An ally to doctors, a diagnostic art,
MRI's companion, a contrast profound,
Unraveling mysteries, where answers are found.

But beyond the lab, your essence extends,
A symbol of resilience, on which life depends,
Gadolinium, you teach us to adapt and grow,
To face the unknown, with an unwavering glow.

In the tapestry of elements, you are unique,
A beacon of hope, a treasure we seek,

A testament to nature's intricate design,
Gadolinium, a wonder, forever enshrined.
    So let us celebrate this element rare,
Gadolinium, with all its flair,
From scientific marvels to poetic verse,
You captivate us, from beginning to traverse.
    In the symphony of elements, you take the stage,
Gadolinium, an eternal source of sage,
A vibrant element, forever profound,
In the world of chemistry, you truly astound.

# TEN

# DAZZLING

Gadolinium, a shimmering delight,
In the realm of elements, you shine so bright,
From azure hues to emerald gleams,
Your beauty dances in spectral beams.

In the world of medicine, you play a key role,
An MRI ally, enhancing the whole,
A contrast agent, a vibrant guide,
Revealing secrets that cannot hide.

But beyond the lab, your essence expands,
A symbol of resilience in life's shifting sands,
Gadolinium, you teach us to adapt,
To face challenges with courage unwrapped.

In the tapestry of elements, you weave,
Gadolinium, a gem that we receive,
A testament to nature's creative hand,
An element of wonder, both rare and grand.

So let us honor your atomic grace,
Gadolinium, in this poetic embrace,
With each electron, a story untold,
A treasure of science, worth more than gold.

In the symphony of elements, you play,
Gadolinium, forever in our hearts, you stay,
A dazzling element, so pure and true,
Gadolinium, we celebrate you.

# ELEVEN

# ELEGANCE AND EMBRACE

Gadolinium, a luminescent star,
In the vast cosmos, you travel far,
A magnetic gem, with powers untold,
Shimmering bright, a beauty to behold.

In the realm of science, your worth unveiled,
A contrast agent, where secrets are detailed,
MRI's companion, painting vivid scenes,
Revealing the unseen, like mystical dreams.

Yet beyond the lab, your essence transcends,
A symbol of strength, where resilience extends,
Gadolinium, you teach us to endure,
To embrace uncertainty, and remain secure.

In the tapestry of elements, you shine,
Gadolinium, a treasure so fine,

A testament to nature's infinite art,
An element of wonder, etched within our heart.
  So let us celebrate your atomic grace,
Gadolinium, with elegance and embrace,
In the symphony of elements, you play,
Gadolinium, forever casting your radiant ray.

# TWELVE

# LIKE NO OTHER

In the realm of elements, rare and profound,
Gadolinium, your presence astounds,
A magnetic marvel, with a vibrant soul,
Unveiling mysteries, a tale to unfold.

In the realm of medicine, you hold the key,
Enhancing diagnostics, with accuracy,
An ally to MRI, a contrast supreme,
Gadolinium, you make images gleam.

Beyond the lab, your story takes flight,
A symbol of resilience, shining so bright,
Gadolinium, you teach us to adapt,
In the face of challenges, we will never be trapped.

In the tapestry of elements, you shine,
Gadolinium, a treasure so fine,

A testament to nature's creative art,
An element of wonder, etched in every heart.
    So let us celebrate your atomic grace,
Gadolinium, with elegance and embrace,
In the symphony of elements, you play,
Gadolinium, forever casting your radiant ray.
    From laboratories to poetic verse,
Gadolinium, you continue to immerse,
Our minds and hearts in awe and wonder,
A chemical element, like no other.

# THIRTEEN

# BOLD AND REFINED

Gadolinium, oh element divine,
A tale of wonder, let me now define,
In the realm of chemistry, you hold your ground,
With magnetic prowess, so astound.

MRI's ally, you lend your light,
Guiding doctors through the darkest night,
Contrast agent, revealing secrets deep,
Unveiling mysteries, a treasure to keep.

But beyond the lab, your story unfolds,
A symbol of strength, as life unfolds,
Gadolinium, you teach us to persist,
To overcome hurdles, and never resist.

In the tapestry of elements, you shine,
Gadolinium, a jewel so fine,
A testament to nature's artful hand,
An element unique, forever grand.

So let us celebrate your atomic grace,
Gadolinium, in this poetic space,
You dance with electrons, a cosmic ballet,
Enchanting us all, in your radiant display.

In the symphony of elements, you play,
Gadolinium, a star that won't fade away,
A chemical marvel, both bold and refined,
Gadolinium, forever etched in our mind.

# FOURTEEN

# FOR ALL TO SEE

Oh, Gadolinium, element divine,
With properties rare, you truly shine,
A magnetic marvel, captivating all,
In your atomic dance, you stand tall.

In the tapestry of elements, you're distinct,
Gadolinium, a gem that's succinct,
A testament to nature's wondrous blend,
An element of fascination, from beginning to end.

Embracing contrasts, you play your part,
As an MRI's companion, you reveal the heart,
Unveiling secrets, like a cosmic key,
Gadolinium, you unlock what's meant to be.

In the symphony of elements, your song resounds,
Gadolinium, a harmony that astounds,
A chemical symphony, composed with grace,
Captivating scientists, exploring space.

So let us celebrate your atomic might,
Gadolinium, a guiding star in the night,
A luminescent gem, forever to be,
An element of wonder, for all to see.

# FIFTEEN

# CAPTIVATING

Gadolinium, a shimmering light,
In the realm of elements, a radiant sight,
A metal of wonder, a treasure untold,
Within your atomic structure, secrets enfold.

In the tapestry of nature, you stand apart,
Gadolinium, a gem cherished in heart,
A symbol of resilience, steadfast and true,
Teaching us to endure, to push through.

In the symphony of elements, you play,
Gadolinium, with elegance and sway,
A conductor of reactions, orchestrating the show,
Unraveling mysteries, as only you know.

From labs to hospitals, your purpose unfolds,
Gadolinium, a healer in stories untold,
A contrast agent, painting vibrant scenes,
Guiding us to answers, where hope convenes.

So let us celebrate your atomic grace,
Gadolinium, a marvel we embrace,
In the world of chemistry, you shine bright,
A captivating element, a beacon of light.

# SIXTEEN

# INSPIRE

Gadolinium, oh element divine,
With a glow that enchants, so sublime,
In the realm of chemistry, you stand tall,
A symbol of wonder, captivating all.

In the tapestry of elements, you gleam,
Gadolinium, a luminescent dream,
With magnetic properties, you astound,
Unveiling secrets, where they're found.

A conductor of change, you illuminate,
Gadolinium, orchestrating fate,
From MRI scans to fluorescent lights,
You bring clarity, banishing the nights.

A guardian of health, you lend your aid,
Gadolinium, a healer unswayed,
Through contrast enhancement, you reveal,
The hidden realms, where illnesses steal.

So let us celebrate your atomic might,
Gadolinium, a beacon of light,
In the realm of science, you inspire,
A versatile element, we admire.

# SEVENTEEN

# CHERISHED AND DEFINED

Gadolinium, a shimmering star,
A chemical wonder from afar,
In the realm of elements, you stand,
With grace and brilliance, hand in hand.

Magnetic properties, you possess,
A magnetic shield, an ironclad dress,
With strength and power, you command,
A force of nature, mighty and grand.

From MRI machines to scientific lore,
You unlock mysteries, forevermore,
With contrast agents, you paint the way,
Revealing truths, come what may.

Gadolinium, a guardian of health,
A healer, bringing hope and stealth,

In the human body, you navigate,
Guiding doctors, shaping fate.

In the symphony of elements, you play,
Gadolinium, a melody that won't sway,
A harmonious blend of science and art,
A chemical masterpiece, a work of heart.

So let us celebrate your atomic grace,
Gadolinium, in this sacred space,
A symbol of wonder, forever enshrined,
A precious element, cherished and defined.

# EIGHTEEN

# ADMIRE YOUR ALLURE

Gadolinium, a jewel of the periodic table,
A luminescent star, forever stable,
In the realm of elements, you stand tall,
A captivating presence, enchanting all.
    With magnetic allure, you captivate,
Gadolinium, a force we celebrate,
Guiding the waves of resonance,
Unveiling secrets, with elegance.
    In the tapestry of atoms, you shine,
Gadolinium, a treasure so fine,
A guardian of health, a healing light,
Navigating ailments, banishing the night.
    Through MRI's lens, you paint the scene,
Gadolinium, revealing what's unseen,

A contrast agent, a beacon of clarity,
Guiding physicians, with unwavering sincerity.
    So let us honor your atomic might,
Gadolinium, a symbol of insight,
In the vast expanse of scientific quest,
You shine, Gadolinium, among the best.
    Forever etched in the annals of discovery,
Gadolinium, a legacy that will forever be,
A chemical marvel, an element so pure,
Gadolinium, we admire your allure.

# NINETEEN

# GUARDIAN OF HEALTH

Gadolinium, a jewel in the periodic chart,
A magnetic marvel, playing its part,
In the realm of elements, you hold your ground,
A versatile gem, forever renowned.

With atomic strength and magnetic might,
Gadolinium, you captivate our sight,
A conductor of spins, a quantum dance,
Unveiling the secrets of magnetic resonance.

In the realm of medicine, you shine,
Gadolinium, a healer so divine,
As a contrast agent, you guide the way,
Revealing the path to a brighter day.

A guardian of health, you bring relief,
Gadolinium, the essence of belief,

In the MRI's embrace, you reveal,
The hidden stories that we seek to heal.
    So let us celebrate your atomic grace,
Gadolinium, a gem we embrace,
In the tapestry of elements, you stand,
A radiant presence, forever in demand.

# TWENTY

# ATOMIC MIGHT

In the realm of chemistry, a gem so rare,
Gadolinium, with an atomic flair,
A magnetic force, you proudly possess,
Unveiling wonders, no one can suppress.

With a shimmering gleam and a vibrant hue,
Gadolinium, you mesmerize our view,
In the MRI's realm, you take the lead,
Guiding doctors, fulfilling their need.

A contrast agent, you paint the scene,
Gadolinium, like an artist so keen,
Revealing the secrets hidden within,
Aiding diagnosis, where hope begins.

A guardian of health, a beacon of light,
Gadolinium, you're a guiding sight,
Through the body's maze, you navigate,
Uncovering truths, erasing the weight.

So let us honor your atomic might,
Gadolinium, a symbol burning bright,
In the world of elements, you shine,
A testament to science, truly divine.

# TWENTY-ONE

# REMARKABLE

In the realm of elements, a gem so rare,
Gadolinium, with brilliance beyond compare,
Your atomic number, a testament to strength,
A symbol of resilience, traversing any length.

Within the MRI's magnetic embrace,
Gadolinium, you reveal with grace,
A contrast agent, illuminating the way,
Unveiling truths, where shadows sway.

Through the body's labyrinth, you guide,
Gadolinium, a healer by our side,
Aiding diagnosis, where answers reside,
Offering hope, with each scan applied.

A guardian of health, a beacon of light,
Gadolinium, shining through the night,

In laboratories, your secrets unfold,
Unleashing discoveries, a story yet untold.
  So let us celebrate your atomic might,
Gadolinium, a symbol burning bright,
In the realm of chemistry, you stand,
A remarkable element, crafted by nature's hand.

# TWENTY-TWO

## ESSENCE OF GRAND

In the tapestry of elements, you're rare,
Gadolinium, a jewel beyond compare,
With electrons spinning in magnetic grace,
You captivate, embodying cosmic embrace.
 A conductor of change, you wield your might,
Gadolinium, orchestrating day and night,
In MRI's realm, you paint vibrant scenes,
Revealing truths hidden within human beings.
 A guardian of health, compassionate and true,
Gadolinium, you heal, our gratitude to you,
As a contrast agent, you guide with precision,
Unveiling ailments, leading us to decision.
 Your atomic structure, a symphony divine,
Gadolinium, resonating through space and time,

Within laboratories, your secrets unfurled,
Unleashing knowledge, expanding our world.
   So let us celebrate your atomic glory,
Gadolinium, a testament to scientific story,
In the pantheon of elements, you stand,
A luminescent marvel, the essence of grand.

# TWENTY-THREE

# CHEMISTRY'S ART

Gadolinium, a jewel in chemistry's crown,
With atomic grace, you astound,
A magnetic force, pulling us near,
Revealing mysteries, crystal clear.

Within the MRI's magnetic domain,
Gadolinium, you bring forth gain,
A contrast agent, a beacon of light,
Guiding doctors, making their insights bright.

A guardian of health, you tirelessly strive,
Gadolinium, keeping hope alive,
In the depths of our bodies, you explore,
Unveiling conditions, seeking to restore.

Your atomic number, a mark of distinction,
Gadolinium, a symbol of scientific precision,

Through laboratories, your secrets unfold,
Unleashing knowledge, stories untold.
    So let us celebrate your elemental might,
Gadolinium, shining ever so bright,
In the vast universe of chemistry's art,
You, Gadolinium, hold a special part.

# TWENTY-FOUR

# SYMBOL OF PROGRESS

In the realm of chemistry, a jewel so rare,
Gadolinium, with brilliance beyond compare,
A magnetic soul, captivating our gaze,
Unveiling mysteries in its atomic ways.

Within the MRI's domain, you reside,
Gadolinium, a guide through the inside,
A contrast agent, bringing clarity and light,
Revealing hidden truths, banishing the night.

A guardian of health, a beacon of hope,
Gadolinium, with healing powers you cope,
Navigating the body's intricate maze,
Detecting ailments, guiding doctors' appraise.

Your atomic symphony, a mesmerizing score,
Gadolinium, resonating at the core,
Through laboratories, your secrets unfurl,
Advancing knowledge, pushing the world.

So let us celebrate your atomic might,
Gadolinium, a star shining so bright,
In the realm of elements, you stand tall,
A symbol of progress, a wondrous marvel.

# TWENTY-FIVE

# ATOMIC STRUCTURE

Gadolinium, an element rare and true,
In the realm of chemistry, we turn to you,
A magnetic marvel, you hold the key,
Unlocking mysteries, for all to see.

Within MRI machines, your power lies,
Gadolinium, revealing truths with no disguise,
A contrast agent, painting vibrant scenes,
Guiding doctors, unraveling what illness means.

A guardian of health, a silent hero,
Gadolinium, your presence brings a glow,
Through the body's pathways, you navigate,
Detecting abnormalities, altering fate.

Your atomic structure, a symphony of grace,
Gadolinium, a marvel in this cosmic space,
In laboratories, your secrets unfold,
Advancing science, as stories are told.

So let us celebrate your atomic might,
Gadolinium, shining with pure light,
In the world of elements, you hold a prime,
A catalyst for knowledge, for all of time.

# TWENTY-SIX

## 64

Oh, Gadolinium, element so rare,
With atomic number 64, beyond compare,
In the realm of chemistry, you hold your place,
A versatile marvel, full of grace.

In the MRI's magnetic field,
Gadolinium, your secrets are revealed,
A contrast agent, enhancing the view,
Guiding physicians, helping lives renew.

A guardian of health, you silently fight,
Gadolinium, shining with healing light,
Through the intricate pathways of our frame,
Unveiling ailments, bringing hope in your name.

Your atomic symphony, a dance divine,
Gadolinium, harmonizing within each line,
In laboratories, your wonders unfold,
Unleashing discoveries, stories yet untold.

So let us celebrate your atomic might,
Gadolinium, a beacon burning bright,
In the vast universe of elements, you stand,
A remarkable treasure, crafted by nature's hand.

# TWENTY-SEVEN

# MAGNETIC GLANCE

Gadolinium, element of wonder,
In the realm of chemistry, you ponder,
A contrast agent, a gleaming guide,
Through the body's depths, you stride.

A guardian of health, steadfast and true,
Gadolinium, we owe our gratitude to you,
Within the magnetic resonance, you dance,
Unveiling secrets with a magnetic glance.

Your atomic structure, a symphony divine,
Gadolinium, orchestrating atoms in line,
In laboratories, your mysteries unfold,
Pushing boundaries, discovering untold.

So let us celebrate your atomic grace,
Gadolinium, shining in this cosmic space,

In the periodic table, you find your place,
A symbol of progress, a source of embrace.

# TWENTY-EIGHT

# CONTRAST AGENT

Gadolinium, a jewel of rare alloy,
In the realm of elements, you bring joy,
A contrast agent, a luminescent key,
Unlocking secrets, enabling us to see.

A guardian of health, a beacon of light,
Gadolinium, shining through the night,
Within the MRI's magnetic embrace,
You reveal the hidden, with utmost grace.

Your atomic symphony, a melody of might,
Gadolinium, orchestrating the atomic flight,
In laboratories, your wonders are explored,
Advancing science, leaving us in awe.

So let us celebrate your atomic reign,
Gadolinium, a star that shall never wane,
In the grand tapestry of the chemical domain,
You stand as a testament, forever to remain.

# TWENTY-NINE

## CELESTIAL LIGHT

Gadolinium, element of intrigue,
In the atomic world, you take the lead,
A contrast agent, a radiant hue,
Revealing mysteries, bringing life anew.

A guardian of health, a silent defender,
Gadolinium, your purpose is tender,
Within the magnetic resonance's domain,
You illuminate pathways, easing pain.

Your atomic structure, a symphony's verse,
Gadolinium, in science, you immerse,
In laboratories, your secrets unfurl,
Advancing knowledge, changing the world.

So let us honor your atomic might,
Gadolinium, shining with celestial light,

In the vast expanse of elements, you reside,
A remarkable force, standing with pride.

# THIRTY

# TREASURE OF SCIENCE

Gadolinium, a precious gem in disguise,
In the realm of elements, you mesmerize,
A contrast agent, a painter's brush,
Revealing hidden landscapes with a hush.

A guardian of health, a silent guide,
Gadolinium, in you we confide,
Within the body's labyrinthine maze,
You navigate, with brilliance ablaze.

Your atomic symphony, a celestial song,
Gadolinium, in harmony you belong,
In laboratories, your secrets unwind,
Fueling discoveries of a brilliant kind.

So let us celebrate your atomic reign,
Gadolinium, a beacon that won't wane,

In the tapestry of elements, you shine,
A treasure of science, forever divine.

# THIRTY-ONE

# ATOMIC REIGN

Gadolinium, a luminescent gem,
In the realm of elements, you stem,
A contrast agent, a vivid hue,
Unveiling truths, revealing what's true.

A guardian of health, a silent knight,
Gadolinium, guiding us towards light,
Within the MRI's magnetic dance,
You paint a picture of our inner expanse.

Your atomic symphony, a cosmic rhyme,
Gadolinium, orchestrating in perfect time,
In laboratories, your wonders unfold,
Igniting innovation, stories yet untold.

So let us honor your atomic reign,
Gadolinium, a force we can't explain,
In the vast cosmos of elements, you gleam,
A radiant star, a scientist's dream.

# THIRTY-TWO

# GADOLINIUM, ORCHESTRATING ATOMS

Gadolinium, a marvel in the periodic chart,
An element of intrigue, playing its part,
A contrast agent, a painter's brush,
Unveiling hidden realms with a gentle hush.

A guardian of health, a silent guide,
Gadolinium, in your presence we confide,
Within the MRI's magnetic embrace,
You illuminate the path, leaving no trace.

Your atomic symphony, a captivating tune,
Gadolinium, orchestrating atoms in the moon,
In laboratories, your secrets unfold,
Fueling discoveries, stories yet untold.

So let us celebrate your atomic might,

Gadolinium, shining through the night,
In the vast expanse of the chemical domain,
You stand as a symbol, forever to remain.

# THIRTY-THREE

## SCIENTIFIC SWIRL

Gadolinium, element of rare allure,
In the realm of science, you endure,
A contrast agent, a celestial prism,
Unlocking mysteries with your atomic rhythm.

A guardian of health, a silent healer,
Gadolinium, your presence is surreal,
Within the MRI's magnetic domain,
You illuminate the path, easing our pain.

Your atomic symphony, a harmonious blend,
Gadolinium, orchestrating elements to transcend,
In laboratories, your secrets unfurl,
Advancing knowledge, a scientific swirl.

So let us celebrate your atomic reign,
Gadolinium, a gem that won't wane,
In the grand tapestry of the periodic chart,
You stand apart, a masterpiece of art.

# THIRTY-FOUR

## STAND UNIQUE

Gadolinium, a shimmering star,
In the realm of elements, you're not far,
A contrast agent, a brilliant light,
Revealing hidden truths, shining so bright.

A guardian of health, a silent guide,
Gadolinium, by our side you reside,
Within the MRI's magnetic embrace,
You paint a picture, with detailed grace.

Your atomic symphony, a cosmic dance,
Gadolinium, in molecules you enhance,
In laboratories, your secrets unfold,
Advancing science, with stories untold.

So let us honor your atomic might,
Gadolinium, a beacon in the night,

In the vast expanse of the periodic domain,
You stand unique, a treasure to attain.

# THIRTY-FIVE

# FASCINATION

Gadolinium, a gem in the periodic table,
Your presence, like a celestial fable,
A contrast agent, painting vivid scenes,
Revealing truths with your magnetic means.
    A guardian of health, a silent sentinel,
Gadolinium, you protect and excel,
Within the MRI's magnetic embrace,
You unveil the secrets, leaving no trace.
    Your atomic symphony, a harmonious chord,
Gadolinium, in science you are adored,
In laboratories, your wonders unfold,
Advancing knowledge, a story yet untold.
    So let us celebrate your atomic might,
Gadolinium, a star shining bright,

In the vast expanse of the chemical realm,
You stand unique, at the helm.
   A symbol of progress, of innovation,
Gadolinium, a beacon of fascination,
In the grand tapestry of elements, you gleam,
A remarkable element, a visionary's dream.

# ABOUT THE AUTHOR

Walter the Educator is one of the pseudonyms for Walter Anderson. Formally educated in Chemistry, Business, and Education, he is an educator, an author, a diverse entrepreneur, and he is the son of a disabled war veteran. "Walter the Educator" shares his time between educating and creating. He holds interests and owns several creative projects that entertain, enlighten, enhance, and educate, hoping to inspire and motivate you.

> Follow, find new works, and stay up to date
> with Walter the Educator™
> at WaltertheEducator.com

www.ingramcontent.com/pod-product-compliance
Lightning Source LLC
LaVergne TN
LVHW052001060526
838201LV00059B/3775